AVOIDING EXTINCTION

Alexander Matthews and
Gordon Ellis

Copyright© 2021 Alexander Matthews and Gordon Ellis

All rights reserved

No part of the book may be reproduced, or stored in a retrieval system, or transmitted in any form or by any means, electronic, mechanical, photocopying, recorded, or otherwise, without express written permission of the publisher.

ISBN: 9798598318980

INTRODUCTION

This book was written by two people with a common background in Philosophy, both of whom are creative writers: one a playwright the other a poet. Two people, who are among many thousands, to share a concern with the condition of the world and the prospect that the highway being taken leads to its inevitable outcome in human extinction, and environmental and planetary destruction.

Even where there is recognition that there could be a problem, attitudes lag behind both the urgency required to address the issues and a recognition that there are a series of interconnected issues that need to be tackled, ranging from our view of what it is to be a human being, social justice, economic policy, food production, the role of science and technology in our life, the nature and functioning of the media, and our currently debased and malfunctioning political system. Climate and environmental issues can not be detached from the broader context that they are a symptom of: a manifestations of a deep-seated dysfunctionality that needs to be acknowledged and addressed to ensure both long-term survival and human and planetary flourishing.

In this short work, we can only begin the process of identifying some of the problems and start to address what needs to be done. It is, we hope, a wake up call (yet another!), and contribution to dialogue. It is time both to act decisively and to reflect deeply. This crisis presents us with a unique opportunity - are we up to taking it and fulfilling our most profound human aspirations or will we go under, or perhaps just about stay afloat amidst the proliferating wreckage?

AVOIDING EXTINCTION

AVOIDING EXTINCTION

You have stolen my childhood. You are failing us. People are suffering, people are dying, entire ecosystems are collapsing. We are in the beginning of a mass extinction, and all you can talk about is money and fairy tales of eternal economic growth. How dare you continue to look away and come here and say you're doing enough when the politics and solutions needed are still nowhere in sight?

(Greta Thunberg addressing the United Nations 2019)

We need a sense of universal responsibility as our central motivation to rebalance our relations with the environment and our neighbours. Appreciating the oneness of humanity in the face of the challenge of global warming is the real key to our survival.

(Dalai Lama, p.34, 'Our Only Home')

Greta Thunberg and other young people have done a splendid job in galvanising public opinion on the idea of human extinction, something that our governments have failed to do in the last 60 years. The problem is that, once that attention is galvanised, where does it turn to? How does it cope collectively with the situation, unless it has some knowledge of where the trouble lies?

The original purpose of this short work was to direct that galvanised attention to the sources of experiments and even productions which would make our future more viable. However, having spent some time researching these projects, we found that (a) there were already experts writing about them and (b) there was far too much ground to cover in a mere paper or even a large book. Also, each time we looked on the internet, the ground had shifted somewhat, so there are lots of viable concerns which are competing with each other and must be encouraged by having attention drawn to them and funds as well. So the purpose of this work has been revamped to suggest mental attitudes which may block or get in the way of this movement towards a viable future. In the final section of this book, we will make suggestions as to where to look and how to take action in terms of a viable future.

But first, below, we will consider obstructive attitudes. Many young people have ignored them and just got on with their work towards a viable future, which is admirable, But, for those of us who get stuck in these obstructive attitudes, it is salutary to name them and shame them. Many of these attitudes are based on relative ignorance. We are pretty sure that, if someone stood up in Parliament and talked, for example, about the dangers of radiation, no MP could describe exactly what radiation is and how it works. Or, if they did, they'd have a very hazy idea. And yet they're being asked to vote on Hinkley Point and other nuclear reactors, and maybe even small nuclear reactors. They are having to decide whether to allow EDF, a French company, to spend 18.9 billion pounds (and this sum is rising exponentially) on building a new reactor when everybody knows that EDF has built two prototypes before and both have failed. It is this sort of crazy situation and others like it that could and should be avoided if people had more knowledge about where the problems lie and what to do about them.

1.

The poet Leopardi[1] contends that man cannot live without illusions, but now we can see that mankind cannot live with illusions, not, that is, if there is to be any hope of continuing existence. The situation is critical, and there is every reason to panic if that is what is required to engage in the necessary vital, urgent action before the last opportunity passes us by. We are sleepwalking into disaster, oblivious to what the meaning and impact of environmental catastrophe amounts to, and in denial about the relationship of that to the socio-economic and political system that has given rise to it, and whose maintenance sustains and accelerates not just the crisis, but our complacency.

Complete destruction of organised human life, the extinction of many species of other creatures, as well as vast numbers of those species that survive, and a near total degradation of the natural world so that it becomes uninhabitable, this is what turning our backs, shutting it out, will bring about; our ignorance forming the basis of our complicity in this act of mass and gratuitous vandalism, driven by greed. The obsession with passing on property and wealth to our children forming an unforgivable preoccupation in the face of the appalling conditions they will inherit, with reducing chances of survival, as a result of our criminal neglect. Leopardi is correct, man does live in illusions, but a condition of his continuing survival is that this should cease to be the case, and that he should face up to the reality of how things are and act before there are only illusions left. There has never been a time when honesty is more important, yet at such a time there has never been such a paucity of it: the lie flourishes like Japanese knotweed, particularly virulent in the political and social domains, not to mention the tales we tell ourselves, the self-deception and denial we conceal ourselves behind. The destruction of life around us, and its support-systems, undermines, and renders impossible, the continuing existence of the human species as well, because everything is interconnected.

Avoiding Extinction

It is interesting that the recent Climate Assembly in the UK[2] did not regard the situation as unduly urgent and seemed keen, overall, to protect lifestyles and leave changes as voluntary. That altered slightly with the impact of Covid-19, in that more began to wonder about travel and work conditions and about its occurrence providing an opportunity to effect a change in lifestyles. What this shows is that unless people are immediately, and adversely, affected by conditions outside their control, they are likely to be resistant to change, that is to say (a) wish it to be up to individuals who want to preserve their lifestyle, (b) not see any pressing need for immediate or drastic action, unlike, for example, in events such as Covid-19 or mobilisation in the event of war. Of course, the Assembly didn't look more deeply into the underlying causes and predicated recommendation on the preservation of the existing system and ideology, unlike Pope Frances in his Encyclical Letter Laudato Si' – 'On Care for Our Common Home'.[3] Interestingly, although the Assembly emphasised the need for education, information, and media utilisation, nowhere was there reference, for instance, to the 'The Earth Charter'[4] or other such documents. There seemed to be an overall air of complacency emanating from their conclusions, which were careful to avoid anything that might be challenging or stirring, and in some ways only proved their point about the need for more education and information, somewhat like the Titanic, with the band playing on, except that it won't be icebergs that are hit but a perfect, and uncontrollable, storm. On the other hand, being educated and in possession of information doesn't seem to be sufficient, as in the case of Andreas Malm of Lund University, who calls for a Communist War,[5] invoking the figure of Lenin, against climate change, probably only marginally less palatable than a call for a Maoist Cultural Revolution. Leftist posturing, and theorising, is not going to stop the rampages of the Right.

Also, looking at Covid-19, for instance, there seemed no recognition by the Climate Assembly of the man-made extent and nature of this disease and the fact that man's activities are creating the conditions for further viruses crossing species and spreading rapidly among the human population to perhaps even more devastating effect. This, too, is something that needs to be urgently addressed and cannot be left to voluntary means. There needs to be a change of worldview, and there also needs to be a legislative change. For instance, there should be an offence of Ecocide,[6] ranking with Genocide as the most serious offences that can be committed. We need to go beyond labelling food sources and leaving it to individual consumers to choose; it needs to be against the law for anyone to sell goods that have any environmentally destructive effect such as have been caused by the extension of palm oil, soybean, beef, wood products, and cocoa production resulting from deforestation and biodiversity loss. And it is vital that infringements are detected and incur very heavy penalties. It should be a criminal offence for banks or other financial institutions, and industries, to lend money, or in any other way aid, the causes of ecological destruction as well as individuals engaged in such pursuits, and this needs to be rigorously enforced. For instance, last year a new report shows that 'The world's largest investment banks provided more than $2.6tn of finance linked to the destruction of ecosystems and wildlife. Among the worst offenders were the Bank of America,

Citigroup and JP Morgan Chase.'[7] Institutions should urgently be required to adopt and enforce policies that protect ecosystems and wildlife. This applies, notably, to areas such as forestry, mining, food production (industrial farming etc), fossil fuels, infrastructure, tourism and transport. This is one category of crime that would fall under the heading of Ecocide. It needs to be, as it was with the abolition of slavery or child labour, absolutely unacceptable. It should also be an offence to express climate denial just as it is with holocaust denial, as the effects of the former are even more appallingly destructive. Unless governments show themselves to be totally serious, and committed through appropriate legislation and a new legal framework, there will not be the degree of galvanisation required to effect the enormous changes necessary, going beyond the reduction of greenhouse gases by 2050, which is only one part of the problem. This has to be done in such a way that ensures the survival (including of other species and plants) of humans, and promotes global justice, which is indivisible from climate and biodiversity issues.

There also needs to be the abolition of the status of, 'Limited Liability Company'; and companies, and their directors specifically, need to be held fully responsible for the effects and consequences of their decisions and actions. To achieve this there needs to be a legal framework requiring consultation, and working with, all interested parties. The idea of the swashbuckling business entrepreneur who overcomes all obstacles to establish dominance, wealth, and status, is dead and needs to be interred in the vaults of history. Further, where wealth has been accumulated, whether by individuals or organisations, as a result of the deprivation, degradation or destruction of others, or environments, that wealth should be appropriated and returned to those who have been exploited and lost out to enable them to invest in a sustainable future. Such practices of banditry need to be outlawed and held as unacceptable business practice. It is simply wrong to have business directors that are the equivalent to being under age or mentally incapacitated and therefore protected from the full legal consequences of their actions.

Avoiding Extinction

There needs to be a fundamental change in the orientation of our Western societies and the ideology that has prevailed, leading us to the present catastrophic situation. Since the end of the nineteenth century there has been an awareness, among thinkers of note, of the magnitude of the crisis confronting us, and the untenability of 'Western Civilisation' as it currently stands. There are a number of factors here, the most significant perhaps being the growth of liberalism / neoliberalism, becoming libertarianism, and the pursuit of unlimited growth – the two being closely related. This has also been enabled by, while further enabling, the cult of the individual and the sovereignty of a subjectivist attitude in the name of what has erroneously been presented as freedom, becoming exemplified in the replacement of the person by the consumer, and of virtue by expediency. What we have lost is balance and integration. A view of the individual-in-the-world-with-others, who is not, and cannot be, independent but is interdependent and where there is no individual that stands apart from, and can simply operate on, a separate, objective world. In other words that dualism, as an outlook, is false and the effects of subscribing to this view and acting on it have been devastating. At a deep level, there needs to be a fundamental change here for any enduring change: a shift to living in an interdependent world where there is concern for the common good. One could say, in current terms, that self-interest lies in the interest of others, and that a failure in this is degradation and annihilation.

The ideas of progress and limitless growth lie very much in nineteenth century England with the liberal ideology of laissez-faire or, in contemporary terms, 'let it rip', which sanctioned, indeed required, the capitalist plundering of environmental resources and exploitation of human beings. There need to be free markets, free labour, to enable continual expansion and growth. This, more recently, resulted in the insulting doctrine of 'trickle-down' economics as a rationalisation to justify an ever-increasing wealth gap and environmental disregard; and extensive financial deregulation, with the encouragement of indebtedness from cradle to grave. It has also meant further cultural impoverishment, the undermining of any effective moral immune system to leave people open to persuasion and exploitation, and narcissistic self-absorption. In all of this, being a *Person* is reduced to little other than consisting of being an unholy trinity of producer, distributor and consumer rolled into one.

2.

Yet there were voices already, during the nineteenth century, raised against the notion of continuous growth, such as that of John Stuart Mill. In 'The Principle of Political Economy', Mill advocates the realisation of a stationary state, as there cannot be a boundless increase in wealth, and such a condition would be a vast improvement on how things were then in the nineteenth century England, let alone today. Coupled with this there should be a 'better distribution of property', and a 'system of legislation favouring equality of fortunes, so far as this is consistent with the just claim of individuals to the fruits, whether great or small, of his or her own industry'. As a result society would have these features:

'A well-paid and affluent body of labourers; no enormous fortunes, except where earned and accumulated during a single

lifetime; but a much larger body of persons than at present not only exempt from the coarser toils, but with sufficient leisure, both physical and mental, from mechanical details, to cultivate freely the graces of life, and afford examples of them to the classes less favourably circumstanced for their growth.'[8]

He says that a stationary state is completely compatible with human improvements. Mill, of course, did not consider the environmental degradation, the threat to sentient survival, and the perpetration of gross injustices among indigenous peoples and in the 'developing world'. The adoption of the radicalised form of neoliberalism has exacerbated an already intolerable situation. We find among its eminent proponents in the public sphere, Alan Greenspan,[9] a former head of the Federal Reserve in the US, a follower of the ideas of Ayn Rand,[10] high priestess of selfishness and fervent advocate of the elimination of compassion. Political life in England and the US, after the duopoly of Reagan and Thatcher with their cohorts and enablers, has been firmly entrenched in this ideology with catastrophic consequences both internally and externally, and has fostered the breakdown of any sense of a common good, given rise to libertarianism and led to the growth of neofascism. It has also been the main engine of ecological and wildlife destruction. It is part of the problem that needs to be tackled, having had a profound influence on those brought up under its doctrines, who have known nothing else; and their attitudes, values and behaviours. It is vital that public life should become a domain where truth, honesty, transparency, and the common good are central. The state has a crucial role to play if the climate crisis is to be addressed, which is not to belittle the changes and contributions that also need to be made at an individual level and collectively within society.

Gregory Bateson,[11] in 'Mind and Nature', gives a vivid example of unrestricted growth when he considers the case of the creation of a Polyploid horse. This horse is twice the size of a Clydesdale, twice as long and high, and twice as thick, with four times the number of chromosomes. As a result it is too heavy to stand unaided and needs to be raised and supported by a crane. It has to be perpetually hosed down to keep it cool, otherwise its insides would cook, the skin and dermal fat being twice as thick while the surface area was only four times that of a normal horse. Although its brain was eight times the normal size, it showed no greater intelligence and it was always panting, both to keep cool and to oxygenate its body. It would eat eight times that of the normal horse down an oesophagus only four times that of normal, while its blood vessels were reduced relative to its size. It was a sorry specimen that was not independently viable without extensive, artificial, life-support.

Whereas in life outside the laboratory and the city there are natural limits to growth, which is controlled by subtle and extensive systems, this fable draws attention to some of the problems of artificial, created growth: growth that is not regulated and kept in balance with environmental factors, but ignores and rides rough-shod over them.

Unfortunately, growth, economic growth, is still what is pursued and is the measure by which a society and an individual are judged, and indeed by which a person judges themselves a success or failure. Creativity is rewarded primarily for new products, or enhancement of existing ones, and huge investment towards this end is made in scientific research and technological application which provide the engines of growth! (An interesting idea, that engines are the means of growth.) This becomes almost an end in itself, with marginal consideration for any effects or implications, and requires a narrowing of focus and desensitising, as with military personnel, and a detachment from the supporting natural environment, including feelings, hence the importance of large institutions that are fortified against extraneous intrusions and thrive on procedures and targets. This is the domain of Western civilisation.

Your civilisation is killing the Earth, is the message that Nemonte Nenquimo, of the Indigenous peoples of the Amazon, recently gave via The Guardian newspaper.[12] She relates that Westerners do not respect the planet and show deep ignorance in their attitudes and actions, as well as an arrogance and disregard for others, as a result of their spiritual impoverishment. They show an incapacity to relate, and are hell-bent on plunder and despoliation with a total disregard for the impact of their behaviour, behaving like bandits and robbers. She contrasts the disposition of the Indigenous peoples towards their environment, which they have learned about and adjusted to for over 1000 years, enabling them to not only survive but to thrive. That is, until the impact of Western civilisation. She draws attention to the centrality of love for the indigenous peoples, in the sense of a profound reverence for the world and for life.

This is a profound and crucial message, calling for a reorientation of our life and our values. It is not enough to just come up with clever solutions within the existent dominant ideology; the very ideology itself is the cause of the problem. That, for example, deregulation, and its close cousin libertarianism, are forms of insanity. It is a quiet message that Nemonte Nenquimo gives us, but is it one that will be heard amidst the flurry of speculation and sea of information (and disinformation), the scramble for the techno-fix that will maintain the status quo?

This attitude is clearly represented by the Independent journalist, John Rentoul, in an article.[13]. In it Rentoul emphasises that any solutions to global warming, specifically with respect to achieving zero-carbon emissions, is predicated on the preservation of the lifestyle of rich countries, and that if it costs too much it won't happen. This is a widespread, if bizarre, notion; basically, maintenance of the lifestyle of a select group of people is preferable to life itself. Rather planetary extinction than that we should forego the comforts and luxuries of an indulgent lifestyle that is parasitic upon the exploitation of others as well as of the planet. Yes, of course, there is a fundamental problem of values that this illustrates. Spiritual impoverishment? – Evidently so. We should even accept nuclear power, as advocated by Zion Lights of Extinction Rebellion.[14] Perhaps people should get a life rather than a lifestyle, while it is still possible.

John Rental seems to believe that the situation is now more hopeful, there is really only an image problem to overcome so that something rosy and sunny can be presented rather than doom and gloom: a job for the advertising executives and sleight-of-mouth politicians. Interestingly, he is more sceptical of China's recent commitment to achieve zero-carbon emissions by 2060. This from someone with near zero knowledge of China and contrary to the views advanced by those more expert in the field, such as Barbara Finamore[15] who makes the point that, 'China has a track record of under-promising and over-delivering on its climate commitments', and is in a strong structural position to do so, unlike, for instance, the UK in the recent distracted expostulation by its Prime Minister – all wind and no turbines! Once a decision is made, the Chinese government has the resolve and the means to implement it, and has shown a will do so with a determination. The record of the current Prime Minister, both as Mayor of London and now, shows bluster, indecision, incompetent decision, inaction followed by wrong action, a proclamation of victory in the jaws of defeat, deceit and deception, none of which would be acceptable even in an advertising executive.

Closer to hand, Germany is another interesting place to consider with respect to attitudes and activities concerning the environment.[16]

Germany, unlike France and the UK, is a non-nuclear state, and one that has shown widespread aversion to the nuclear industry, despite its having successfully lobbied for construction of 17 nuclear plants.[17] It has been fought every step of the way from the first proposal in early 1975 at Wyhl on the French border, which was abandoned with the site being turned into a nature reserve, up until the final decision to phase out all nuclear energy by 2021. En route, there were the threats posed by the Tihange reactor at Huy in Belgium, the impact of Chernobyl in 1986 where Germany was in direct line of the spread of radioactivity, necessitating emergency action, and the devastation of Fukushima which had a profound impact on German public opinion. When the German government tried to extend the running time of nuclear plants beyond 2021 there was widespread opposition, with three-quarters of voters being opposed to it.

Events of the summer of 2018 also had a deep impact on the German people, bringing the climate crisis literally home. It is one thing to know about rising sea levels, wild-fires, excessive temperatures and lack of rainfall, but these remain somewhat abstract as long as they are occurring elsewhere and have little immediate impact. That summer, however, there were very high temperatures in Germany, crops were ruined, rivers dried up and cargo traffic on the Rhine halted for the first time in living memory. Industry was also adversely affected and had to scale back production and find innovative ways of continuing to function. The clincher, however, was the forests ablaze; forests which German people have an almost primal relationship with. This provided a profound shock to their psyche. As a result in 2019, the government introduced the far-reaching Climate Protection Law[18], which critics argue is still quite inadequate for the task in hand. Indeed, corporations, the coal and car industries in particular, are held to have had an undue adverse influence on the outcomes. Germany has over 1000 non-governmental researchers working on environmental and climate issues, with a similar number in universities. So the issues are being taken very seriously. There is opposition, not only from vested interests in maintaining the status quo, but also from the political right wing, the AfD and libertarians. Despite the strong Green movement, with the Green Party shaping up to be political power-brokers, and maybe even gaining the role of Chancellor in the near future, there are still enormous problems that need to be tackled and there is a major question over whether Germany will be able to reduce CO_2 emissions by 40% by 2030. What has been evident is that it is only in the face of a current crisis that people have mobilised to bring about effective change, and that in a population that is generally well disposed to climate and environmental issues. Here again, though, there is no question of deep changes, only in techno-fixes to enable the current lifestyle to continue and be further enhanced.

3.

We have seen, when considering the situation in Germany, the important role played by emotion in making decisions and effecting action. Indeed, it is possible to engage in a rational decision making procedure, identify alternative courses of action, and be unable to make a decision precisely because of the absence of emotion – that which moves, which provides and sustains motive. Thus we can have an abundance of 'knowledge', or information, that makes no difference whatsoever. Often the problem is too much information (and control by vested interests), and also the wrong kind of knowledge, which constitutes a form of ignorance. It takes wisdom to decide what is appropriate, and wisdom is precisely something which is not valued and is lacking. Our capacity to perceive clearly can be obscured by surfeit as much as be lacking through deficit. And in the absence of feeling and vision it can amount to very little, or else misperception.

In the case of Germany, and indeed the West generally, it would appear that the dominant emotions driving climate and environmental action are anxiety and fear, which are reactive and in a way disengaging or defensive and negative. This contrasts markedly with the orientation expressed by Nemonte Nenquimo, which is one of love and reverence, and hence caring. This is a mode of engagement and is positive. As such it becomes self-sustaining and meaningful as a way of life whereas anxiety and fear are corrosive and nihilistic, giving rise to desperation and despair. Nemonte Nenquimo's is what could be called a spiritual or religious disposition embodied in a way of life that makes life inherently worth living. It is not enough at all to simply look at the 'problems' and marshal forces to overcome them like a form of military endeavour. This could be seen as a form of violence and a manifestation of the same kind of mind that gave rise to them in the first place.

4.

We ask the question, 'What would a viable future be?' because there is a problem with how things are currently and the thought that they are not sustainable. We need to consider how they are, and what it is about them that renders them problematic, so problematic in fact that they cast a future into doubt. We also need to reflect that our present now was also once a future for those who preceded us, a future that most will not have lived to see.

When we look back, there was never a time that was without problems, so it is not that having problems is unique to our time. There have been droughts, floods, plagues, wars, social problems, environmental threats, and the effects of psychopathology[19]. The manifesting of problems has challenged humans to seek means of overcoming them, frequently off-setting short-term gain with longer term pain in the form of more acute problems that have needed to be addressed. What is it that makes our current situation so different?

That we are in a crisis, or multiple possibly terminal crises, would seem to be widely accepted from those with substantial quantities of information and data to those who just have a sense that there is something deeply wrong, even where they do not wish to face up to it fully, and are in denial, perhaps mainly because it constitutes a threat to their way of life, which they do not want to change, or be changed against their will. There is a loosening of any sense of coherence or, indeed, purpose, informing the life of human beings, with an increasing sense of anxiety and desperation. As it happens, this suits the dominant ideology, driven by economic theory and practice, as it renders individuals susceptible to exploitation and increased profits for a few. Indeed, it is important that the individual does not have any means of resistance, has strong inducements, and is isolated, and hence vulnerable, and in need in such a way that large scale economic production can pose as the means of satisfying those amplified needs.

One major difference today is the scale and range of problems, and the fact that they affect everyone on the planet, in differing ways and to varying degrees, and not just people but other forms of sentient life, as well as plant life and ecosystems upon which life is dependent. The projected trajectory also points to an intensification and compounding of problems and increase in their magnitude. Social dysfunction, and individual and relationship breakdown, are endemic and increasingly evident, as well as the commercialisation of an array of means of treatment, as objects of consumption. Indeed, if one considers the situation of the social body as a whole, it is hard not to conclude that it is riddled with disease, which is spreading increasingly rapidly, in circumstances that are conducive to its flourishing, and what we see around us as an ecological crisis are the manifest symptoms of this disease. In terms of addressing the issues, of course there is a need to alleviate the effects of the symptoms, and this requires practical measures and substantive changes, but that is not enough without addressing the causes. It can only buy time, and delay, only to possibly even intensify and assist the spread of the disease subsequently, as has been happening. The scale of what is involved, and its extent, requires fundamental change with the removal of, at least, means of controlling the disease itself, which most likely cannot be eliminated entirely. A two pronged approach is thus required – emergency measures to mitigate the worst symptoms to prevent the spread and development of the disease's symptoms, and addressing the causes of the disease, and the circumstances that allow it to flourish and spread, introducing the means to allow health and flourishing. Central to this is the notion of what a human being is, what their place is within the scheme of things, and how such beings can flourish.

The tendency is to identify particular problem areas such as: transport, energy, food, housing, nature, and communications and then look for specific solutions.

Invariably, these are technological solutions which we reach for because they are 'to hand' and presuppose that how we live needs to be sustained albeit in modified form, one where technology has an increasingly dominant role to play. There is a kind of colonisation of life by technology to the increasing exclusion of those things that are quintessential to the manifestation of life with the obvious, logical conclusion of the elimination of life completely. This is not seen at all, or seen to be so distant, or not plausible (which seems to involve a degree of denial and wishful-thinking) and therefore not to be a matter of concern. There are also those who would welcome such developments, leaders in the field standing to benefit enormously financially and in terms of status and power. There is thus a kind of compulsion, an impulsion, to persist along the path of what has been termed 'progress', with either no particular goal or some technological utopia.

This excludes consideration of the roots of the problem which lie 'within' rather than 'outside', with the outside changes conditioning perception and understanding to constitute further barriers to looking for root causes rather than quick-fixes-with-continuity, and assimilating human thinking and disposition further to machines, albeit soft ones. The effect is increasing threats of dehumanisation or alienation and recourse to a variety of means of attempting to deal with this or capitulating. For many this is just going-with-the-flow. It is rather like the Buddhist story of the king who, being forewarned that a rain would fall the drinking of which would turn people insane, refrained from drinking the rainwater only to find himself isolated and ridiculed for his oddness. As a result he drank the rainwater in order to be like his fellow men and to communicate with fellow human beings. In current circumstances such assimilation, or connivance, simply contributes to consolidation of the disease and ensures its spreading. So either one does that, withdraws and waits for the flood-waters to rise, or tries to find a way of communicating that enables others to reflect and come to act differently, effecting a change of values and life.

This is not to say that action can be delayed in addressing multiple crises. This is required urgently; however it is only the means of keeping the patient alive and in a relatively stable condition so that further actions can be taken addressing the root causes. Otherwise we simply continue stumbling from one problem to others of even greater magnitude. We need to address, as I said, the question of what it is to be human and how humans can best thrive, and this also embraces the question of politics, economics and religion, the *umwelt*[20] or milieu in which humans live. Without a consensus on these, within which there will inevitably be variations, there is neither the coherence nor the means of acting that are necessary for unified and concerted action, and nothing less than this is required where survival is at stake, let alone thriving. If we have a misconception of what it is to be human, or no overt conception at all, then what follows will be a travesty and misalignment, and this, at least, has been what has been happening. We need to change our way of thinking.

So first, perhaps we need to consider the practical steps that are required to be taken as a matter of urgency to ensure the patient does not expire. This includes the establishment of a legal framework that has Ecocide as a specific criminal offence, together with auxiliary offences, and a means of vigorous enforcement. There need to be regulations on financial institutions and lending, and on corporations, particularly requiring them to engage in definitive consultation with all relevant parties to reach a joint decision by all who stand to be affected. There need to be restrictions on what can be imported and exported, and what can be sold within a country, relating to the nature of the production and its process, sources, and ethical standards. There needs to be a Global Green New Deal as proposed by Robert Pollin[21], and others, ensuring justice for those who stand to lose out as a result of the changes required, recognising their human dignity and right to work, and actively making provision for it during the transition to a Green Economy. All products need to demonstrate clearly that they can be either recycled or reabsorbed into the ecosphere without any deleterious effects. We need to aim at zero-waste as well as zero-carbon. There needs to be social justice and economic equalisation ensuring that there are no great disparities of wealth, both within a country and between countries. Military thinking, and finances, but not military means, need to go into defending mankind from his greatest enemy: the climate catastrophe that he has launched in a process of insane self-destruction. The 'Earth Charter'[22] speaks clearly of the need to:

Demilitarise national security systems to the level of a non-provocative defence posture, and convert military resources to peaceful purposes, including ecological restoration.

The 'Earth Charter' seems to be an excellent and succinct basis to orientate from and to disseminate widely within schools and throughout society providing distinct guidelines for action. It is also important that corporate lobbying and influencing should be excluded. There needs to be transparency, so anything they have to say needs to be made public and open to challenge. Their role and influence has been a major source of profound problems and needs to be radically reduced, while retaining a voice, one among others, and no greater. Smaller dedicated businesses and cooperatives need to be encouraged. The fossil fuel industry, and its auxiliaries, need to be closed, with provision for workers currently employed, and new job prospects. Travel needs to be reduced, especially by plane, and economies weaned off dependence on tourism. In conjunction with this there needs to be a redesign and renewal of local urban environments. The Victorians left a legacy of public parks, and since then little has been done to 'green' and make sociable overcrowded urban space and support communities. Communities should be engaged in effecting urban transformation to enhance the quality of life and opportunities.

Other measures to change hearts and minds need to be begun at the same time, I would suggest, to ensure there is no relapse or continuing dependence on destructive means of living, so that addictions are addressed and a new view of life and ways of thinking considered.

5.

Scientific investigation – processes and procedures – can be dangerous and pose a high risk. Side effects, or auxiliary consequences, can be unanticipated or unprepared for. Additionally, any context of origination, and application, of (pure) research is regarded as irrelevant or value-neutral.

How research is to be used, why it is being carried out, who is promoting it and utilising it, are regarded as superfluous considerations – little to do with science which is hermetically sealed having almost nothing to do with human beings or society. Science's concern, it is said, is to provide an explanation, a truth, concerning the world as an independently existent reality (assumption) and how it works.

We are used to detaching science from our evaluative life as humans. It is therefore detached from the life-world, just as the scientist is detached from any values or feelings. It is pure intellectual excitement, augmented by empirical evaluation where possible and appropriate. How the world is to be is irrelevant to pure research, and extraneous to the narrowly defined purpose of applied research which is part of the economic and technological structures. That such-and-such will result, or can be achieved, can be ascertained, but its value and ramification are irrelevant. This is highly problematic. More recently, some scientists have questioned this viewpoint, cf. Carlo Rovelli[23], also Alexander Matthews.[24]

6.

There are a number of obstructive and negative attitudes which get in the way of avoiding extinction.

The first sort of attitude is mainly among scientists and perhaps academic researchers. This attitude values knowledge as higher than human safety and even survival. You could suppose it is understandable because in a sense knowing is part and parcel of surviving, so the mindset is rationalised that way but nevertheless extremely dangerous. We can see this in the Cassini[25] and twelve other planned subsequent launches all involving a payload of plutonium. Critics suggested that there was a ten per cent chance of failure at launch. This could have wiped out, or made uninhabitable, large parts of Florida and there would, of course, have been far more pollution involving the southern states of the United States. Considering the half-life of plutonium to be hundreds of thousands of years, this would have been catastrophic. Nevertheless, scientists went ahead with the launching and thank goodness it proved successful. The production and storage of plutonium is another example. According to Sipri reports the UK has the world's largest stockpile. It has been estimated that all humans have a tiny fraction of lethal plutonium in their reproductive organs. Those living in the Northern Hemisphere have more.

There are a number of further examples of this sort of mindset, which is highly dangerous to human survival. To go into these in detail, please see 'How Some Scientists Erode the Human Rights We Value, by Alexander Matthews. 26 Here is a list of further examples of this dangerous mind set taken from that article. refusing to take inadequate computers off nuclear alert even though they are not what is called ' Y2K' compliant.In layman's terms: inadequate for the task. Research into the Big Bang. The Euratom Treaty involving death by small doses. Soil tests condemning Sellafield.[27] Dumping nuclear waste. GM foods and intended Monsanto monopoly. The genome project. Refitting Trident at Plymouth. Tinkering with the Van Allen Belt.[28] All of these have put humanity in grave danger and yet they have been pursued in the cause of science. This mindset is perhaps the most direct threat to extinction rebellion. The rebellion part would be very much against this mindset. There are also huge monopolies to consider and militarising space when there is too much space junk up there already in the stratosphere and it is too dangerous for astronauts to get through, making manned flights extremely hazardous. The media do not report these scientific ventures on the whole or, if they do, it would be in some obscure part of a paper like page 26 or a localised journal. Most of the young that are now in the know eschew newspapers and get their information off the internet. Hopefully there may be enough young people who work for a viable future to overcome this type of obstructive mindset. It is after all perhaps the most direct threat to the extinction of the human race. It is also involved with iffy moral decisions like human cloning, and whether humans are so much like machines that they will become machines.[29]

The second obstructive mindset has to do with governments. Governments tend to look for short-term solutions. They make purely economic decisions, not evaluative ones. And it's evaluative decisions which involve extinction rebellion. Not only that but their decisions are usually based on national interests rather than global ones and the problems involving human survival are nearly all global. This has allowed politicians to ignore human extinction for the last 60 years, despite the fact that the problem is getting more and more serious. The popular media does the same. And there are D notices to manoeuvre around. Again, this attitude has to be overcome by those who seek to enhance a viable future.

Then there is the monopolist mindset whereby people who start and are in charge of large monopolies expand monetarily and again make purely financial decisions rather than evaluative ones, i.e. in biblical terms, worshipping the golden calf. These financial decisions invariably help the company involved and thereby impoverish everybody else. As Buckminster Fuller once put it so eloquently, if the money were shared fairly and equally among the citizens of a country (I think he was thinking of the US) everybody could be a millionaire. Everyone. So those who are involved in extinction rebellion would have to overcome this mindset.

Every morning we open the newspapers and there are the people who are robbing us staring us straight in the face from these large monopolies.

The fourth mindset – and we are sure there are plenty of others – involves older people who simply say, well, 'I'll be dead and it's therefore not my problem'. This in fact may be wrong. Armageddon has arrived in the form of a pandemic and flooding, increased storms, forest fires, ice melting and so forth. This reminds me forcibly of Yeats's great poem 'The Second Coming'[30], *the best lack all conviction and the worst are full of passionate intensity.*

A fifth mindset has to do with the media. This is simply not dealing with the problem by ignoring it. It could have to do with D notices or considerations over what sells newspapers, etc. Too often, the news simply reports what has happened and not why. For example, it hasn't connected up the pandemic with our attacks on biodiversity. Luckily, other commentators such as David Attenborough have. In any case it is a short-term view and really not worthy of newspapers generally and it could contribute to the reasons why newspapers are having trouble selling enough. Their editors should be responsible and take a long-term view and join the extinction rebellion. We invite them to do so.

There are a number of other types of obstructive mindsets, but these will have to be overcome in order to deal with some of the problems we have. Luckily we think there are enough young people who are getting the idea that we're in real trouble and more and more are joining forces to avoid extinction every day, which is encouraging to say the least. Hopefully their combined efforts will be enough to offset these obstructive and suicidal mindsets.

These attitudes, collectively or by themselves, can deprive the younger generation of a future tense. It would be humanly impossible to lead anything like what we call life in two tenses, the present and the past. We function as sentient human beings by gleaning from the future predictions that enable us to live as rational and happy human beings. We need three tenses to live. But, if we take away the future, that deprives the human race of the means to live a fully sentient human life. So, psychologically as well as physically, it is important to zero in on these problems in order to suggest that there is some sort of viable future.

Happily, most areas of concern are being worked on already by companies who hope to make their fortunes by break-throughs in their research. In the meantime, some have run into problems with what they are up to and they may need money. What we are suggesting in this essay is that those who are interested in a viable future and want to do something about it, i.e. they don't have the above mindsets, pick on one or, if they have the energy, more of these issues, and do research into what is needed and try to raise money for them. If you do this, you will be doing your bit for a viable future.

Here, briefly, is a list of some of the concerns we now have:

Avoiding Extinction

Perhaps the most difficult problem is overpopulation. We are over seven billion people using seven times the amount of resources that the earth could safely supply. After all, we have only one planet to live on.

Then there is transport. Planes spew out contrails, which have to do with global warming and they must look for a better fuel. The same goes for ships, or storing electric batteries for cars

Too much concrete and other factors interrupt the nitrogen cycle, which is necessary to human life.

Pesticides destroy bees and bugs and other living creatures.

There are companies working on a number of ways of decarbonisation. I think, on the internet, about thirty are listed.

In regard to housing, each new house should be regarded as far as possible as a mini power station , using renewable energy.

There are problems with palm oil and deforestation.

There is a company working on turning plastics into oil. This company is called Carbon Engineering.

Hydrogen is a fuel source of the future. If we can protect ourselves against its high flammability.

For those of you who want a general overall picture, there is an excellent article by Lord Rees entitled, 'Our world faces many perils but we will prevail.'[31]

There is research into the Van Allen belt and militarisation of space, projects which should be discouraged in that they threaten our fragile ionosphere.

The seas need our protection, not just from plastics but from over-fishing, from chemicals dumped in the oceans.

We must be sure that cloud cover doesn't become too prevalent from jet trails, that the Gulf Stream is not polluted.

Deforestation must be stopped, and more trees planted. Trees are the lungs of the earth.

Air quality must be improved.

Water quality must be improved.

All these and many more are projects for people interested in the future to work on. In the process, as we say, they will probably encounter all sorts of obstructive attitudes but they will also be encouraged by the number of projects and enterprises and even some solutions towards a viable future.

There is more than enough renewable energy to be garnished from the atmosphere. People will feel that solar energy by itself is not enough, but it can be transported via fibre optic cables with a three per cent loss over ten thousand kilometres. Large sections of the desert could be reclaimed as food growing land under these solar panels.

If those of you who are interested spent just two per cent of your time doing this very important work, this would be a wonderful boon for the future. For all the plants and the animals that need us.

In previous campaigns over the last fifty years, I (Alexander) found that, as I get sucked into the research involved and the writing it up and campaigning, more and more of my time is taken up. That is when I tend to ease off but perhaps you can be more disciplined than that or simply delegate research to those other people who are interested. As evidence of people's current attitudes and behaviour the following report gives a not unreasonable indication for where things stand at the moment.

7.

The 'People and Power Report 2020,'[32] was based on a survey of 2000 people in October of this year. It was produced by an organisation called Pure Planet. The report's headline was, 'People want to be more sustainable, what's stopping them?' Overall it found that people believe in sustainability, and the vast majority believe they could do more. However, among the 18-34 year olds only 11% are doing all they can to take action, while among the over-55s only 27% are and 17% have done nothing to reduce their impact on the climate. Switching from oil or gas heating, buying an electric car and switching to a renewable energy supplier are particularly poorly supported. Yet, over four fifths of those interviewed believe we have a responsibility to prevent climate change and three quarters say they try and reduce their environmental impact where possible. Unfortunately, too many don't see this as being possible for them. Reasons given for this are: cost, effort, sacrifice, so it is either too expensive to live sustainably, it requires too much of this time or effort and, although they want change, they are not prepared to make sacrifices to achieve it. Now, there are problems with cost, for instance, although it may be cheaper to run an electric vehicle they are much more expensive to purchase. There is also uncertainty about battery life and recharging points, given the reduced milage that has been possible, although these factors will no doubt change. Take-up of new technology is often slow because the initial products are high-end and expensive, and the technology regarded as untested over time, so this is not surprising. There is also the need to ensure battery recycling. There is thus been an aura of uncertainty and prohibitive outlay with respect to electric vehicles. To switch to alternative energy for homes also requires a much greater initial outlay for biomass boilers, Ground Source Heat Pumps and Air Source Heat Pumps. They are also unknowns for most people, who would rather stick with what they know, and may be sceptical about company hype. Also there is little support, for instance, for domestic biomass boilers whose local service engineer could be over a hundred miles away, and whose service contracts, and replacement parts are prohibitively expensive compared to a standard gas or oil boiler. There is also the additional problem of having to shift 25kg bags of pellets after delivery, and for daily use, which requires considerable strength and application. The alternative of storage silos which can be filled by blowing the wood pellets into them and then channelled into the house are too big for the average domestic property, and an additional layer of expense. Ground Source Heat Pump and Air Source Heat Pump systems are cheaper to maintain, and require less effort to run, but are still expensive to install and not suitable everywhere. Switching to a renewable energy supplier for electricity, however, need not be more expensive when combined with other means of reducing energy usage, which there is an overall benefit from doing, yet over 40% have no interest in doing so. Among the 18-34 year olds, 21% say they are too busy, 11% just don't want to, and 11% have no interest, just 11% say they are doing all they can (whatever that consists of!). Among the over-55s 27% claim to be doing all they can, very few are too busy or don't want to, although the same percentage 11% have no interest. When it comes to making sacrifices to protect the climate or save the planet, most people reject any measures that impact adversely on their freedom to do what they want to do, on their lifestyle, or on their convenience, as the poet, Charles Olson put it at the beginning of his poem, 'The Kingfishers'[33]:

What does not change/is the will to change.

Thus, a switch to road pricing where you pay for the number of miles you travel, limitation on the number of flights a person can take, additional tax on diesel vehicles, and pollution reduced zones near schools and hospitals, have very little support, while things like increasing investment in renewables, onshore wind farms, subsidised solar panels and banning single use plastics have support ranging from 33% to 40%, and this no doubt would be less if it meant economic adjustments elsewhere that would reduce their spending power and their lifestyle, or otherwise be inconvenient. Although around two thirds of young people, and about half of all people, are more likely to spend money with a business that displays its greed credentials, there is very little trust in any of the agencies that would be responsible for change: around a quarter trust the government and the same for energy businesses, with about the same number not trusting anyone at all. There is marginally greater trust in high profile campaigners, and a negligible amount in the media, household and Tech businesses. As far as preventing climate change is concerned 80% of the over-55s and 69% of

18-34 year olds accepted that we have a responsibility to prevent climate change, and yet among this latter group there is much less actual practical support in doing so, in, for instance, shopping locally, buying fewer clothes that last longer, avoiding single-use plastics, shopping seasonally. On the other hand a greater number say they would support a green recovery (although it is unclear what would be meant by this), have considered buying an electric car, and would be interested in moving into an eco-friendly house in the near future. Only a third of them thought that individual action couldn't make a difference to climate change compared with nearly 60% of those over-55.

Many young people claim to be too busy to do anything to prevent climate change (though busy doing what is not stated). That would elicit responses such as these from the Danish philosopher Soren Kierkegaard:

To be busy means, divided and scattered (depending on the object which occupies one), to occupy oneself with all the manifold things in which it is practically impossible for a man to be whole, whole entirely or whole in a single part, something only a lunatic can successfully do. To be busy means, divided and scattered, to occupy oneself with what makes a man divided and scattered.[34]

And

...in the midst of busyness, double-mindedness is to be found. Just as the echo dwells in the woods, as stillness dwells in the desert, so double-mindedness dwells in the press of busyness. That the one who will the Good only to a certain degree, that he is double-minded, that he has a distracted mind, scarcely needs to be pointed out.....

The press of busyness into which one steadily enters further and further, the noise in which the truth continually slips more and more into oblivion, and the mass of connections, stimuli, and hindrances, these make it ever more impossible for one to win any deeper knowledge (of himself). The idea of being too busy to engage in actions that would prevent a cataclysm sounds either too ridiculous or too callous to entertain. Yet, people can state this without embarrassment, a clear indication that something is profoundly wrong.[35]

As is so often the case, people want things to be different but they don't want to change. There is an understandable hesitancy about technological changes that may appear to be untested, are expensive, and poorly supported, if that involves individual investment. But whereas Bob Dylan, in 'Gonna Change My Way of Thinking'[36], sings of, changing his way of thinking and getting a different set of rules, most people don't want to change their lifestyle and what they see of their personal freedoms, which are actually only those bestowed by the market for the market's gain; even at the cost of the lives of other beings, human and non-human, and the despoliation and destruction of the natural and supportive environment. They wish to stay on, and enhance, the same trajectory, eliminating whatever might prove a barrier to that and creating conditions, including the development of what they hope will be technological solutions at whatever price, in terms of the progress of humans, if necessary into machines.

8.

There are two fundamental problems: the primacy that is given to the supremacy of the individual and the glorification of the self, and a consequent myopia that reinforces the culture of consumption, of grasping and addiction. Underlying both the development of the problems, and sustaining them, is the question of worldview, and the notion of what it is to be human.

There is a very interesting and important article by Lynn White, Jr. that was published in 'Science' journal in 1967 entitled 'The Historical Roots of Our Ecologic Crisis.'[37] And although one can sympathise with the sentiment expressed by Bob Dylan, it is perhaps unfortunate, if understandable given the mainstream culture in the US, that the move should have been towards Christianity, and perhaps especially Protestant Christianity, which stands implicated at the very centre of the catastrophic developments that have produced the existential crisis that Western civilisation has bequeathed to the world. This view is also articulated by White, who himself speaks as a Christian:

We would seem to be headed towards conclusions unpalatable to many Christians. Since both science and technology are blessed words in our contemporary vocabulary, some may be happy at the notions, first, that, viewed historically, modern science is an extrapolation of natural theology and, second, that modern technology is at least partly to be explained as an Occidental, voluntarist realisation of the Christian dogma of man's transcendence of, and rightful mastery over, nature. But, as we now recognise, somewhat over a century ago science and technology - hitherto quite separate activities - joined to give mankind powers which, to judge by many of the ecologic effects, are out of control. If so, Christianity bears a huge burden of guilt.[38]

White also correctly argues that whether or not someone regards themselves as a Christian, or a post-Christian, including all of those who reject religion completely, it is the Christian perspective and values that underpin and inform Western civilisation still. The idea of transcendence, both of a deity and, reflecting that, of man from nature and of man, having named the animals, having dominion over them and the natural world to dispose of in his own interest: here we see the glorification of the self as species - mankind. The view of a fundamental duality, of subject and object, of divine and mundane, of matter and spirit is pervasive and deeply problematic, as are the extreme responses evoked in an attempt to resolve the philosophical problems that duality gives rise to: the reduction of everything to either materialism or physicalism, on the one hand, and idealism or spiritualism on the other. The Judeo-Christian tradition has also bequeathed a linear notion of time in which the key notion becomes progress which, as White points out,

> ...*was unknown either to Graeco-Roman antiquity or to the Orient.*[39]

Elsewhere, although there is the notion of linearity as an unfolding, it takes place within an overall cyclical movement, and of cycles within cycles which reflect to movements in the natural world that man still sees himself as being part of, a natural world that is, moreover, teeming with life, with spirits, to whom he has to establish proper relations, or experience adverse consequences. In essence, there was a great sensitivity to the environment and to proper relationships and balance within it. This is a sensibility with much greater sensitivity and perceptiveness, with the focus on relationship rather than the individual. For Christians, salvation was of the individual, or the essence of the individual, their soul. This was either the gift of their god or something the individual had to strive for, and was the fruit of their own endeavours. This latter came markedly to the fore with the establishment of the Protestant church, with its focus on the individual's own interpretation of the Bible and the sidelining, or exclusion, of tradition or other intermediaries, and within that of those who held that worldly success reflected the degree of their righteousness, and hence eligibility for salvation, and vice versa. This produces the 'self-made man' and, of course, fosters all of those qualities that such material success thrives upon, not least of which is ruthlessness, together with calculation and manipulation.

9.

Michael J. Sandel, in his book 'The Tyranny of Merit',[40] traces the process and the effects of this, from Luther's revolt against the purchasing of merit by the rich to gain admittance to heaven, as offered by the Catholic church through the sale of indulgences, and what was seen to be the accumulation of merit through engaging in rites and rituals, to his proclamation of salvation entirely through the grace of the god, and then to Calvin's doctrine of the predestined elect giving rise to the notion of salvation through successful work, we can trace the emergence of the primacy of the individual. As Sandal puts it with respect to the solution to the question that came to be posed, 'How can I know that I am one of the elect?' - 'Well, by means of the signs, namely through having achieved worldly success. This is an ostensible indication that I am a member of the elect.' This provides a spur to accumulate capital, though within that context, without conspicuous consumption, and following an ascetic mode of life. Whereas, it had been monks in monasteries, who had previously devoted their lives to spiritual cultivation, particularly of more 'inward' qualities, now it was worldly success enjoyed by the businessman or entrepreneur that was a mark of spiritual development and to be lauded. As a consequence:

Avoiding Extinction

This ethic unleashes a torrent of anxious, energetic striving that generates great wealth but at the same time reveals the dark side of responsibility and self-making. The humility prompted by helplessness in the face of grace gives way to the hubris prompted by belief in our own merit.[41]

Over time, the ascetic side became eroded, or discarded, and the view emerged that success and wealth are a sign of merit, suffering, of idleness , with the sanctification of winners and denigration of losers, as Sandal puts it. Gratitude, generosity, and humility are thus regarded as misplaced, and those who fail do not deserve, or warrant, support or the provision of health care. They are, in the effect, the damned. For the enormous number of evangelical prosperity- gospel followers in the US, their god rewards faith with wealth and good health. Indeed, he is an important investment opportunity, because if you give money to him he will return it manyfold. In the words of Joel Osteen, head of the largest church in the US located in Houston, quoted by Sandal:

Jesus died that we might live an abundant life.[42]

And a poll in 'Time' magazine revealed that getting on for two thirds of people is the belief that their god wants them to be prosperous. Only a lack of faith will stop this from happening, which will also be the cause of them suffering ill-fortune and ill-health, and therefore not deserving of any social or health-care provisions. The focus of life is the individual, and their striving, their success, is seen as purely a result of this, with no other factors playing a role. Each person is self-made, independent, and the sole author of their destiny. What happens to others is their problem. This ethos, so prevalent in the US, has successfully migrated across to the UK, spearheaded by Margaret Thatcher and her fervent free-market supporters. They are still a powerful force in the government, and the land, and quite ruthless in the pursuit of their own ends; as fanatical and zealous as any religious fundamentalist (which they are not disconnected with,) and all that enables that regardless of whether anyone else's interest or well-being is affected, after all, as Thatcher proclaimed, 'There is no such thing as society'. And all manifestation of collectivity and community were to be broken-down, while self-concern was to become the central guiding force in a person's life. This fed in to a receptive strata that had been established in the 1960s and 70s with the era of hippy culture, with its focus on self-gratification, indulgence, and narcissism, under the banner of 'Feel the Love'. Self-growth, self-development, self-concern and self-absorption became the preoccupation, which evolved into the preoccupation with therapeutics and New Age beliefs and behaviours. All essentially libertarian, the social dimension of the liberalism promoted free-market economics. Thus the supremacy of the cult of the self, and individual pleasure, at any cost (you can get away with). Psychotherapy has further promoted the focus on 'Me', in so far as it has predominantly advocated that all emotional problems are intrapsychic,[43] and have been preoccupied with creating a map of a fictitious inner world and its dynamics. In Jung's case what he called the quest for individuation, and the attainment of Christ being, was presented as an updating of, and replacement for, the Protestant Christianity he was born into and became disenchanted with, complete with its own priesthood of individuated therapists over which he presided. Interestingly, Jung, although he showed an interest in Oriental religions, rejected a role for them in the Occident, as does Lynn White, forgetting that Christianity itself is an Oriental religion in origin, and was imposed on the majority in the West by means of violence, for instance with Charlemagne; something which proponents, or practitioners, of other non-Abrahamic religions have never sought to do.

Capitalism, as a development out of Protestantism, thrives on the atomisation of a society into individuals, and requires the removal of all barriers to their exploitation, and elevation of all means of promoting addiction, including what is cynically marketed as 'individual freedom'. Science also, through its method of analysis, has been about breaking down and down, in a search for the basic building blocks, which are individuals, while Darwinism as has become popularised, particularly in its Social form, has the slogan, 'survival of the fittest', promoting the idea that it is the fittest individual, in competition with others, who will come out on top and survive, thus sanctifying competition between individuals as a basic human characteristic. When one looks closely at the evidence, this is a complete travesty that has had disastrous consequences. The general developmental course for a person in the West is seen as a growth from dependency to independence. The same is applied to nations, so individual sovereignty is seen as the only means of maintaining an individual identity, as if identity is some kind of core being rather than a process of identification that is engaged in.

There is no core being, no essence, no substance that is independently existent[4], that makes a thing or person what they are, and independence is not the end point but a stage on the way to be superseded by interdependence, which is a coming back to source and recognising how things actually exist. Only on the basis of this recognition can one act appropriately, rather than out of ignorance, and avoid suffering and the proliferation of more and more problems which there is a tendency to address by means of force or violence. It thus becomes possible to live in accord, in harmony, in mutual recognition and support.

So, underlying the enormous changes that are required to enable survival and thriving, is a fundamental conceptual realignment, away from the primacy of a fundamental essence[44], a substance, as well as some absolute, unchanging, and eternal mode of being, to the recognition that the actuality is *relational* and that everything exists interdependently, rather than independently, and is undergoing continuous change, the rate of which is variable and so may not always be detectable, giving things the appearance of permanence. This is an illusion though. Only *abstraction*s can be permanent, in this sense, in so far as they exist apart from the world, and are formal in nature and 'dead'. They are intellectual constructions.

Because every thing is interdependent, and only a temporary constellation, we need to care for all which, where they are also sentient, means we should engage compassionately, recognising that what I am, and become, as a result of my relations with others, from the earliest relation with my mother, with my peers, and those I respect and seek to emulate. I cannot be self-made but am a joint-project, built up over time, and devoid of essence that gives me an identity (other than as a hominid). The 'I' that emerges, through narrative and self-narrative, is what is identified with but is not an entity hidden away in some inaccessible part of myself that remains the same, and indeed will survive the destruction of the body. This does not mean that there can be no spirituality, that everything is purely material or physical; on the contrary it enables a much more profound sense of spirituality that emerges once the fixation on an essential I is abandoned, and everything that has coagulated around it released. Some have experienced this within the Christian tradition, such as St Francis of Assisi and Meister Eckart, both of whom have been condemned as heretical (although Eckhart's death, prematurely, was fortuitous for him). Other religious traditions have also produced people of profound insight and enlightened sensibility, although they have more frequently been on the margins, or rejected by their mainstream tradition. This has happened in spite of the doctrine and dogma, although the experiences have often been presented in the imagery and using terms which are distinctive to that tradition, unsurprisingly. There is still much to learn from them. As Proust said, and a similar sentiment was also expressed by Pope Francis in his wonderful and powerful encyclical, 'Laudato Si", written in the name of, and invoking, St Francis:

...so it is not the activity of the present moment but wise reflections from the past that help us to safeguard our future[45].

10.

The question has frequently occurred: 'What is man?', 'What is it to be human? The answer to this question establishes the context and means by which men and women can seek to live, and what their values are.

Is man just an animal? is he a machine? Is he an embodied soul or spirit? Is he just a social and/or an economic product? Is he a cosmic accident? A random occurrence? Or did he emerge on purpose to fulfil a particular cosmic task? And so the questions can go on, and although it is instructive and fascinating to explore this is not the occasion to do it. We can, however, begin with some basics. In particular, we need to begin with locating the species 'mankind' within the natural world.

Anthropos - mankind - is an integral part of the natural world from which it emerges and which sustains it. This mode of being is classified as Hominid or, more generally, 'Great Ape', so mankind is a member of that family. The distinguishing feature is the greater cognitive capacity that has been elaborated. There is nothing separate, or apart from, this manifest that constitutes the most general reference of the word 'world', or, as Wittgenstein put it, 'The world is all that is the case.'[46] Though the meaning of this requires elucidation, let us use this at the moment.

A man or woman is born into this world and is not yet human, anymore than a chimpanzee or gorilla is human, but becomes human through initiation into a Culture, which consists of, at least: language, institutions, modes of organisation, attitudes and beliefs, and forms of artistic expression.[47] These are an elaboration on the basic forms or themes that are already present in other Hominids, and which are continuous with other life forms. Consciousness, which can be seen to be present in Hominids as well as shared with other life forms, therefore cannot be related to the greater cognitive capacity of humans which is more akin to machines such as computers. It also means that such machines do not require and, qua machines, do not have consciousness. Also, the brain, or specifically the left hemisphere,[48] which is associated with computation and sign manipulation, is not directly related to consciousness and is not its source or cause. It would be more appropriate to relate it to what has traditionally been known as the 'heart', the locus of emotion and intuition, rather than the 'mind' of cognitive processing.[49] Nowadays, this is related to the functions of the right hemisphere of the brain.

Further, there is nothing 'outside' this world that creates it, there is no 'maker', rather the world is self-generating, evolving out of and returning into itself, and consciousness arises within it as an aspect of it.[50]

Categories such as 'heaven' or 'paradise' and 'hell' are culturally dependent phenomena; a culture forming a world within the World, this World thus consisting of many worlds each of which constitutes a view of the World. There is no view of the World outside one of these views, which are human elaborations and projections, constituting the milieu in which humans have a form of life and a world in which to live. These worlds constitute the three dimensions of being human: body, speech and mind, where 'mind' is understood as 'heart' rather than the cognitive processes which are associated with speech. Also, categories such as 'material' and 'spiritual' constitute concepts that are means of organising experience within a particular world, the nature of the experience itself escapes these categorisations. Preceding and underlying this is a unitary experience of a single unfolding reality of which we are part.[51] Whether categories are apt or not depends upon their consequences, in particular whether they strengthen survival and enhance fulfilment. Attempts to reduce everything to just one category to the exclusion of the other, and maintain that as a result everything that exists is just one nature or substance, whether material or spiritual, fails to recognise that the categories are not the reality, simply a means of ordering experience. Theirs is an attempt to create another, monochrome, world, and as such is a cultural product of which we can ask is it enriching? Is it inclusive? From which power-structures does it arise and serve? Overall -

does it provide a sense of coherence? 'Coherence' being the crucial notion, as something experienced, and articulated as well as possible. In particular we need to know whether such actions, following from, are life-enhancing or destructive. A world (system) may not be sustainable, and its incoherence may be obscured, or denied, for a range of reasons, even though the effects of actions, lending towards destruction, become increasingly apparent. Just as mental incoherence is incapacitating in an individual, so an incoherent world-view tends towards disintegration of that world, and of the World that it is a feature of.

Within the Bible, man is said to be made in the image of the god. This is not the case with the bonobo, the lizard or the dung beetle, or any other creature. Man's position is different, unique, set-apart; and it is one where he is credited with dominion over all the Earth, so he has the power and control. Dominion would be understood on the model of the absolute ruler, the potentate. There is thus an original cleavage, as man is separated out and established over-and-against the natural world, mirroring the domination of man over woman, who is there to serve and to fructify. Where lacking, it was man's job, then, to extend and establish domination where it might be lacking and to use and exploit in accordance with his benefit. It is this fundamental model that underlies Western man's being-in-the-world, it is still culturally active even where it has been doctrinally ignored or denied. There is thus Man and there is the Other, and even though this Other has been held to be made by the god, who determined the nature of each of its constituents, it is Man who is given control over it for his own ends. This dualism was deepened by Descartes who rejected the notion that animals could be conscious and were other than machines, as indeed, was man's own body. It is with him that cognition - thinking - becomes identified with consciousness, or being, and a division, and gulf, between mind and body, spirit and matter, has become axiomatic for those who have followed. A characterisation of how things are within a given culture has come to be definitive for how things actually are, shaping how they are experienced and interacted with. Operationally, this has proved catastrophic in its outcomes. Despite which, people will continue to subscribe to it as an article of faith, held as fervently as with any dogmatic religion; and where they have encountered intellectual problems they have sought to resolve them within the framework established by it rather than questioning the framework. Often status, income and vanity have demanded it despite the manifest ruin created, which can be dismissed as not being theoretical and therefore nothing to do with them.

As hominid, humans live in the natural world from which they emerged and which they are dependent upon, but qua humans they arise from and manifest a cultural world which mediates their thinking about the natural world. This latter is a co-construction, a collective artifice, although it is a given to someone born or initiated into it; and although it is an artifice it is experienced as being real. However, compared to the natural world it is an illusion, including the construction that constitutes the natural world as experienced from within the cultural milieu. Although it may be termed an illusion, that doesn't mean it is inconsequential. On the contrary, for humans, and the natural world in so far as they are inescapably part of it, it appears more real and more immediate, as human understanding and action is seen, understood, and acted upon, in terms of this milieu, the concepts and categories that it has given rise to. To be 'human' is to be a cultural construction, but this is simply a medium through which the natural world, that he arose from, comes to be experienced and understood. A problem arises when humans lose contact with the immediacy of their roots as embodiment in a non-artificial world, and fuse their identity with the cultural-product. Increasingly the cultural-product has come to be seen in terms of man's elaborated, abstract, cognitive functions, particularly those associated predominantly with the left hemisphere of the brain. Daniel N. Stern,[52] and others, have shown that the functions of this hemisphere are later to develop than those of the right hemispheric capacities. The elaboration of left hemispheric functions has proceeded independent of the natural world, and have increasingly come to be seen as constituting a totality in themselves that are most purely exemplified in computational machines which can then be embodied in robots. This area has become the focus of investment and development at the expense of those capacities rooted predominantly in the right hemisphere which manifest expressively rather than cognitively. A reason for this is the former provides greater power, control, and economic generativity, the last of these being pivotal. This has led to a downgrading, and disregard, not only of right hemispheric capacities, but also the natural world which has come to be seen as an object and instrument to be used, exploited and finally exhausted and destroyed. A consequence of this is that to survive, if that is any longer the appropriate term, an entity has to cease to be organic, and relational to the natural world, which excludes humans as life forms, as well as the natural world that they are part of. We see the triumph of the rational artefact that only requires an artificial environment for continuity and operation. Such entities require neither feelings nor consciousness, which would be hangovers from an obsolete and eliminated natural world. There would be no purpose in their existence, which would be completely meaningless, and constitute the final triumph of Nihilism. This prospect is what confronts us.

11.

Problems that arise today are rooted in the separation and increasing gulf between the natural and cultural worlds, with the latter struggling to establish itself as autonomous at the expense of the former. This is evident in man's seeing himself as unique, special, separate, and dominant over the natural world, especially through the utilisation of his elaborated cognitive capacity; something which constitutes an impoverishment of his own nature as well of the natural world he is a part of. Cities, megacities, provide a particularly favourable habitat for this way of thinking, and large institutions and organisations are also fertile environments for its flourishing.

Whereas within a religious culture a key concept is that of 'salvation', within a capitalist industrialist culture it is 'consumption', and the life and values adhered to are orientated towards that goal. In the latter the holy trinity are: calculation, manipulation and control, and society is geared towards promoting consumption addiction which it depends on for its survival in that particular form.

Unfortunately, it is self-destructive, sawing off the branch on which it is sitting, as well as eliminating everything else around it in the process. One has to say, it is questionable whether it will be possible to wean humans, within a culture dominated by such an ideology, off this addiction when most neither see where they are nor want to give it up. At the same time while humans have been getting drunk at the bar, so to speak, the scientific-technological forces of abstraction and control have been marched on, becoming a substitute for 'religion' with the hope of salvation placed in them. It is in these that faith is placed, even at the cost of the destruction of the natural environment. After all, they say, we will be able to construct entirely artificial environments that will support intelligent forms, even if human cognitive faculties have migrated to hardware and the human body been discarded, together with everything else that they arose from and had their place among. This is the new promised land, and the new mode of salvation, human beings having become obsolete, it is not they who will be saved, however, but they who will be sacrificed.

We have seen this movement towards abstraction in the arts as well, in painting, music and also in poetry, shifting increasingly away from contact with human embodiment-in-the-world-with-others, and experience as such, to purely formal and mathematical structures that are free-standing, and language becoming detached from articulation of meaning to become free-floating linguistic play and language arrangements, almost as an affirmation of the logical positivist credo: a life devoid of content, which is purely formal, conventional, cut-off from a deeper sense of life and experience.

12.

A central notion of 'coherence' is important here. 'Coherence' is the name we give to the way that things hold together, forming a pattern that is intelligible enabling us to make sense of them, thus giving them meaning for us. Sometimes that coherence may be evident, because of our familiarity with that particular mode or ordering, sometimes not, either through lack of familiarity or degree of complexity or density.

Avoiding Extinction

We can say that for something to come into existence there is a search or movement towards coherence, and that the attainment of coherence is what sustains its existence. There would appear to need to be a process independent of our naming of it and articulating or delineating it. The only way of maintaining incoherence is by means of violence, it cannot be self-sustaining but tends towards decline and disintegration. Coherence is dependent upon mutuality, or interdependence, where the parts hold together to form a whole, itself part of other wholes. The coherence of any one whole is also affected by the coherence of any whole of which it is a part. In other words, nothing is self-sufficient. In any natural organism, there is a movement towards coherence, and where this cannot be achieved, or is interrupted and cannot be rectified, the resulting incoherence gives way to disintegration. This seems to be a natural movement built into the unfolding of existence helping to provide a relative stability. In humans we seek to mirror, in one way or another, this coherence in order to make sense of things and give stability to our lives. We are also inclined to impose an order on it through the structuring of abstract ideas, with the intent to manipulate, dominate and control in a move to fix things down, to make them invariable, beyond change, permanent - but this rigidity or frigidity is a denial of the natural unfolding of the coherence of coherence overall, a breaking off of a part, resolving it into a whole, and determining it once and for all, not dissimilar to taking something from the natural or cultural world where it had a role and a life and installing it in a museum in a glass case which is temperature controlled.

What we need to do is to explore the means of establishing this sense of coherence, for the groups, society, culture, and for the individual, and in such a way that it doesn't come to constitute a straightjacket but a means of enabling and enhancing life as a whole. This involves understanding mankind's place in the world, his relation with other beings around him, and how to live as part of, not apart from, the world that we all share.

13.

When considering the climate emergency, and questions of ecology, it is vitally important not to just address the effects, which are so manifest, but to go beyond that to identify the causes, or grounds, from which the prospectively terminal problems arise.

Of course, there needs to be urgent action to address the effects, such as a Global Green New Deal[53] (because the problems are not merely local or national), a strong and just legislative framework with strictly enforced laws, including international agreements, the wise use (and development) of technology, in accord with, and in support of, life on the Earth and planetary ecology, deep changes in how people live and work with the necessary social and economic readjustment. These are things we have touched on in this essay, but we have also raised questions concerning the causal basis for the manifold problems that constitute a crisis in our civilisation. These need further and deeper discussion giving rise to a fundamental changes in worldview and values if human and other sentient beings and their habitats are to survive and thrive. There is a lot that militates against this - a moribund system, a complacency and apathy that is endemic, and unreflective immersion is a dominant ideology that is pervasive and corrosive. This needs to be challenged if it is to change, it will not do so if left to its own devices. This is a crisis different, and greater in magnitude, than any we have ever faced before and, as things stand, we are set to lose, with catastrophic effects. The extent of the changes needed shouldn't be under-estimated, and will not happen by default.

Unfortunately, the government, or state, is not widely trusted to deal with these issues, and part of the problem is to do with the ideology promoted by governments in England, and elsewhere, that favour vested interests - corporations, financial institutions, big business, including the media, with education posing a different kind of problem, but one not disconnected from the other two. Democracy is distorted by the disproportionate influence of members of those groups the whose wealth, power, and position give them an influence completely disproportionate to their numbers. Ruling parties are susceptible to the promotion of their interests in a way that they are not to the public in general, and not unusually embark on an advertising campaign, utilising distortion and lies, to win the public over to something that may well not be in their best interest, and without people being given the opportunity to evaluate evidence rather than succumb to propaganda. The media further supports these interests, and plays an important role in ensuring that the influence of this powerful minority prevails, through generalisation, distortion and omission, processes which are characteristic of mental dysfunction. Perhaps, in order to turn the tables and ensure that it is the majority and not this powerful minority who prevail, we need to consider something like having a temporary wealth and property disqualification with respect to voting rights, until there is greater equality and justice within society, otherwise this will not happen, and national decisions will favour those vested interests. This would simply be reversing the situation before the 1832 Reform Act whereby only the wealthy could vote and participate in government at all. They could still be free to participate by divesting themselves of excessive assets. Whatever method is decided there needs to be a means for rebalancing society, with less wealth and power discrepancy[54] and greater social justice. This is an issue integral to tackling the climate and biodiversity crisis. We would like to turn now to consider the position of the mass media, which plays such a central role in the life of modern societies in influencing and shaping public opinions and reinforcing dispositions.

14.

The mass media is like a series of mirrors held up to reflect the different interest groups that are operating in a society. Each of the mirrors itself arises in dependence of a number of these groups, for instance: owners, editors, readers, politicians, journalists, advertisers; and each mirror is a manifestation of a different configuration, with politicians being pretty constant throughout. In addition we now have the electronic media, including social media, which are owned by huge technology companies operating universally with the absolute minimum of controls on what is disseminated, either in terms of content or quality, and unanswerable to anyone, unlike the traditional mass media which at least have bodies that can consider and adjudicate complaints. Among these, the BBC, as a public broadcast organisation, is different in having a Trust that happens to be approved by the government in power, appointees invariably reflecting the government's political agenda, so hardly independent or impartial, let alone necessarily committed to public broadcasting. The interplay of the dispositions, agendas and prejudices, of these different constituents gives rise to the phenomenon that we encounter as the mass media. There are other constituents that have been omitted here, including magazines, films, journals, books, billboard hoardings and such like. All of these are operating in, and therefore influenced by and aligned with, the Market whose values and methods extend their tentacles throughout, determining what will work, so what is acceptable, and what not. The Market is the final arbiter of taste, while the government communication department is the source of propaganda of the dominant ideology, and enforcer of that ideology in practice. This is the context of the 'freedom of the media, as indeed it is of freedom of the individual. Within this, as it stands, there is no, or little, means, of establishing and effecting the truth, which is not considered a priority, certainly not compared with power and control. Governments, as such, have stopped governing as their priority and become ruthless ideological machines serving the interests of a particular group in society whose interests they promote, and who exert dominant influence and control within the mass media. Whereas governments, of all political persuasions, sought to govern by consensus in the interest of the country as a whole up until 1979, since then, what has been referred to as the New Right, or the Neoliberals, have sought to re-engineer society in favour of an ideology that favours only a minority and has been ruthless in its pursuit. As part of this process they have re-designated the consensual as Left wing and sought to put relentless pressure on it in, for example, the BBC. Thus the idea of a Common Good has become heretical for the true believer in the New Right Conservatism and every effort has been made to eliminate it in institutions and organisations, with the active and enthusiastic support of the predominantly Right wing media, whose owners and operators stand to gain, together with their friends and associates. There has been a systematic and near-relentless process of undermining any sense of society, solidarity, unity, one nation or country, and the exploitation of fragmentation and atomisation. The press have been a crucial part of this process, owned by vested interests, often not domiciled in the UK, working with, and manipulating, the readership by means of graphic and emotive imagery, over-simplification, and wilful disregard of balanced evaluation. A few brave journalists, and broadcasters, have at times called out the lies, deceits, and revealed the facts that are being traduced, only for organisations employing them to come under increased pressure to exclude them or risk denunciation and marginalisation. All this is done while trumpeting freedom, and at the same time undermining it. Clearly governments cannot be trusted with the media, nor can their communications machinery be trusted. There is ample, compelling, evidence of the chicanery of governments, whose concern is with ideology and the retention and extension of power, and not with governance on behalf of all the people, of the Common Good, let alone planetary well-being as such.

Avoiding Extinction

At the same time, we have the social media where there is license to spread whatever you like, with whatever motive, for whatever end. Like a swamp it can breed disease among people who often have no resistance, appealing to baser elements and to wishful thinking, devoid of any checks and balances. It does nothing to promote reflection, evaluation, or critical engagement which can often happen in face-to-face encounters with other people. Here, subjectivity reigns supreme: 'It is true because I believe it. Because I say it'. Utterances are self-validating, self-evident, truths, and everyone has the right to their own truth, or to share in mine as long as they don't think about it. It also provides a means of collecting information about individuals, and of seeking to influence and manipulate them. In some ways it is the fulfilment of the Capitalists dream, the achievement of their Promised Land. No concern with the Common Good here, or with considered discourse, just a free-for-all.

It is young people mostly who prefer the digital media as sources of information. Among 16-24 year olds 66%, 25-34 year olds 67%, 35-44 year olds 47% and then declining by around 10% for each subsequent band with only 11% of over 65s relying on digital media$_{55}$. Of course, newspapers also have a digital platform, and overall figures appearing in Media Alliance from market research by Nielsen Scarborough for the respective readerships are: Print 53.8%, Newspaper website 41.4%, Mobile 38.6%, Facebook 41.3% and the Huffington Post 39.4% (figures from 2017). Given the decline in the sale of print editions, all the indications are of an increasing trend to on-line sources of information, often of an extraneous and unedited kind.

Of course, newspapers have to appeal to a specific readership, and surveys are carried out to identify what these are so that papers can target them, as can the advertisers whose revenues they are dependent upon. A key factor here affecting not only content but also layout, presentation, and the kind of language used, is reading age.

Reading has been in decline, as reported in the Guardian in February 2020, where it is said that only one quarter of all under-18 year olds read each day. The average age of newspaper readership is 9 years old, with the

Guardian on 14 years and the Sun on 8 years. Readers of tabloids require a smaller vocabulary than those not reading newspapers at all, it has been reported (BBC 6/11/2014). According to the findings Professor Alice Sullivan says:

The presence of tabloid newspapers in the home during childhood was linked to poor cognitive attainment at age 16.

Circulation figures for October 2020 (Press Gazette) have the Daily and Sunday Mail as by far the biggest circulation newspapers, followed by free papers, the Metro and Evening Standard, then the Mirror and the Express, the latter with around one-third of the circulation of the leaders. Readership of the Observer, Guardian and Financial Times, which require a higher reading age, have only a very small fraction of the circulation of the leaders and about a half of that of the Mirror. The Sun and Times, owned by Rupert Murdoch and tax-exiles, the Barclay Brothers, are all absentee owners, living outside the UK. They are keeping their figures secret.

All of this suggests that the overall readership of newspapers has a low reading age and limited vocabulary, and the content and manner of presentation of information has to be shaped accordingly in the circulation wars between competing newspapers. This cannot but affect the standard and quality of information presented as well: it has to be simple, graphic, brief, emotive and require limited cognitive skills, rather like your average pop song. The nature and quality of information given to the readership is the basis for reinforcing prejudices, and views, that further the survival of the newspaper and promote the vested interests of the ownership and the corporate class. This constitutes a major problem to informed, intelligent, public debate and the promotion of the Common Good.

The core of the problem is that we cannot trust in self-regulation in a babel of competing voices seeking to seduce and enthral their audience. Nor can we place confidence in advertising with its sole concern in promoting its products and everything that advances that. We cannot trust in governments to regulate or indeed, as things stand, in their dissemination of information. And newspaper proprietors who seek to exploit, and often inflame, their readership for their own ends, and who do not even inhabit the country where they are selling their 'news', look only to the economic and political benefits of themselves and their friends in positions of wealth and power. Leaving it to market forces and a lightly regulated environment, supports abuse and exploitation, with a frequent disregard for values of integrity and honesty. On the other hand the danger with regulation is the promotion of vested interests and suppression of anything that could expose questioning of those interests by curtailing the press's investigative powers, which have played a vital role within our society, although much less in evidence now, unfortunately. Crucial here, again, is the notion of the Common Good, informed by compassion, integrity and a respect for truth. Advertising requirements that ads need to be Legal, Decent, Honest and True could usefully be applied to news content and to politicians speeches, and breaches truly held to account (of course there are going to be errors, which need to be acknowledged as a minimum). If people are to participate fully in a democracy, it is vital that reading skills and ability to assess information, are vastly improved. So once again education for life, and for democracy, are important, as well as revealing the machinery behind the operation of vested interests for all to see, in which sections of the press have played a vital role, but which needs to be more widely disseminated and available for all. The BBC, as a national institution has played an important role here, but not one beyond reproach, as for example in its coverage of the conflict between Israel and Palestine in the Middle East. Of course, it is not possible to get it right all the time, and the key thing is to learn and adjust through feedback, but not through the pressure of governments, politicians or corporations. The notion of balance is an interesting one also, where, for instance despite the overwhelming, reputable, evidence for climate heating and environmental devastation, the BBC has sought to 'balance' this with climate deniers. It does not, however, 'balance' programmes, or news, on the monarchy with that of Republican views. It is important for such a service to remain vigilant and resistant to pressures from political ideologues determined to impose their agenda, even though apparently respectable and dressed in suits, and to remain sceptical. Political pressures, particularly from the New Right, have been especially problematic and demand resistance if the Common Good is to be sought, which it needs to be, if human and planetary survival and thriving is to be accomplished.

It is abundantly clear that we need to have more, and better, education, and here the problem is again the state which promotes education as a method of training workers, through a process of reducing pupils to little more than robots or learning machines. Of course work, active participation in society through contribution and reward, is an important human need, but a human being is more than a mere functionary, and human qualities transcend mere task completion. Here there is much to learn from approaches outside the existing frame, for instance Steiner and Montessori, and the importance of working with the whole person being-in-the-world with others, with less focus on competition and more on co-operation, the importance of the natural world, and the arts, which are, after all, distinctively human whereas calculation is not. Children also need to be equipped with the means of evaluating, as well as to wonder, so that they will not just accept what is presented unquestioningly, or not examine what is presented. As Socrates said:

The unexamined life is not worth living.

We need to move away from a dualistic philosophy, with its pugilistic orientation, towards a non-dualistic cosmology, with its sense of an unfolding and enfolding whole. This is something that one finds, for instance, in sources as diverse as Classical Taoism/Chan as presented in the translations and works of David Hinton, and in the physicist, David Bohm[56]. Finally, we need to re-orientate towards the centrality of the idea of the Common Good, away from that of the supremacy of the individual with their infallible subjectivity. This is not to say that the individual doesn't matter, and that individual shouldn't be respected, loved, and felt compassion for, on the contrary, it is in their likeness, through what is common, namely the humanity that we share, and the sentience that we share with other creatures, that our life is lived and that we are dependent upon: that we are all interdependent and unique in our particularity, but not set-apart from, better than, superior to others. For Thatcher and the New Right, as exemplified in Ayn Rand, compassion is a dirty word and the c-word that counts is competition. This has been lent a dubious credibility by popularised scientific theorising around the notion of the selfish gene, which came to exemplify what was claimed to be the inherent selfishness of human beings, based on a misconception of what it is to be human (as we indicated earlier), and misapplication of genetic theory. In its damaging and erroneous pursuit Richard Dawkins has displayed the fervour and blind dogmatism of the worst religious zealots that he castigates. It is humility, rather than arrogance, that is a virtue. It is our way of life that is having the devastating consequences that are now becoming so evident, and ultimately need to be addressed and changed. As the poet Rilke said:

For here there is no place
That does not see you. You must change your life.
(Rainer Maria Rilke, 'Archaic Torso of Apollo')

Avoiding Extinction

NOTES

1. Leopardi in his Notebooks (Zibaldone), Penguin Classics, 2013

2. Climate Assembly UK. Climate Assembly Report UK 2020, www.climateassembly.uk

3. Laudato Si', 'Our Common Home' Papal Encyclical. Pope Frances at Vatican Website

4. Earth Charter, www.earthcharter.org

5. Andreas Malm. 'Corona, Climate, Chronic Emergency', Ch.3, Verso, 2020

6. See: Stop Ecocide: www.stopecocide.earth

7. Greenfield (P) and Weston (P): Banks lent $2.6tn to ecosystem and wildlife destruction in 2019 (The Guardian 28/10/20)

8. J.S.Mill, Principles of Political Economy, (Bk 4.3.) p1227, in John Stuart Mill: Ultimate Collection, Madison & Adams Press 2017

9. Alan Greenspan. 'In the early 1950s, Greenspan began an association with novelist and philosopher Ayn Rand. [54] Greenspan was introduced to Rand by his first wife, Joan Mitchell. Rand nicknamed Greenspan "the undertaker" because of his penchant for dark clothing and reserved demeanour. Although Greenspan was initially a logical positivist,[62] he was converted to Rand's philosophy of Objectivism by her associate Nathaniel Branden. He became one of the members of Rand's inner circle, the Ayn Rand Collective, who read Atlas Shrugged while it was being written.' Wikipedia. See also entry in Encyclopaedia Britannica.

Subsequently, some fervent 'Objectivists', came to regard Greenspan as deviating from the purity of Ayn Rand's ideology, an issue which is contested.

10. Ayn Rand, who said: 'Altruism is a moral theory which preaches that man must sacrifice himself for others, that he must place the interests of others above his own, that he must live for the sake of others. Altruism is a monstrous notion, it is the morality of cannibals devouring one another. It is a theory of profound hatred for man, for reason, for achievement, for any form of human success or happiness on earth.' (The Sanction of the Victims' 1981. A lecture at the National Committee for Monetary Reform in New Orleans). See The Ayn Rand Institute, www.ayrand.org

and

'Capitalism and altruism are incompatible; they are philosophical opposites; they cannot co-exist in the same man or the same society'. (Conservatism: An Obituary', 1960) (Ibid)

In contrast, the Dalai Lama, who has presided over numerous dialogues with scientists over many years, says:

'Scientists have concluded that basic human nature is compassionate. Those who grow up in a more compassionate atmosphere tend to be happier and more successful. On the other hand, scientists suggest that living with constant anger or fear undermines our immune system. Hence, compassion and warm-heartedness are not only important at the beginning of our lives but also in the middle and end.' ('Our Only Home' pp.46-7 Bloomsbury Sigma 2020).

And The Care Collective in, 'The Care Manifesto', p.4 (Verso 2020):

'After all, the archetypal neoliberal subject is the entrepreneurial individual whose only relationship to other people is competitive self-enhancement. And the dominant model of social organisation that has emerged is one of competition rather than co-operation. Neoliberalism, in other words, has neither an effective practice of, nor a vocabulary for, care. This has wrought devastating consequences'

It is not accidental that someone like Alan Greenspan moved from Logical Positivism, whose British advocate A.J.Ayer proclaimed that all moral, psychological, social and aesthetic talk was strictly meaningless (see'Language, Truth and Logic') to embrace 'Objectivism' as advocated by Rand. Nor is it accidental that in the field of science, which has increasingly sought to colonise all other domains, it has insisted on its own particular form of objectivism being the only significant one. (See Thomas Nagel, 'The View From Nowhere', Introduction and Ch. 2, OUP, 1986). As a philosophical view, 'Objectivism' has become dominant within society, and the damaging effects of this are evident all around us. But neither is the extreme of 'Subjectivism', which some have retreated into as a result, the answer.

11. Bateson (G), Mind and Nature, pp. 66-7, Fontana/Collins 1980.

12. Nemonte Nenquimo: 'This is my message to the western world – your civilisation is killing life on Earth'. (The Guardian 12/10/20)

13. Rentoul (J): Is Boris Johnson's dream of a carbon-free future realistic? (The Independent 20/10/20)

14. Zion Lights. Referred to in 11.

15. Finamore (B): What China's plan for net-zero emissions by 2060 means for the climate (The Guardian 6/10/20).

16. 'Why the Germans Do it Better', John Kampfner, Atlantic Books 2020

17. Ibid. Ch 7 'No More Pillepalle', p.247ff

18. Climate Protection Law, June 2019. Ibid. p.267

19. Psychopathology: 'psychopathy' is defined as, 'a persistent disorder or disability of mind (whether or not including subnormality of intelligence) which results in abnormally aggressive or seriously irresponsible conduct on the part of the patient..'(Mental Health Act 1959). See 'Critical Dictionary of Psychoanalysis' , Charles Rycroft, Penguin Books 1995. It is proposed that a group of people can be regarded as equivalent to an individual in terms of characteristic behaviour.

20. 'Umwelt': from Jakob von Uexküll. As used in ethnology it is 'an organism's unique sensory world'. (Encyclopaedia Britannica)

21. Robert Pollins in 'Climate Crisis and the Global Green New Deal', Ch. 3. By Noam Chomsky and Robert Pollin, Verso, 2020.

22. Earth Charter: www.earthcharter.org

23. Carlo Rovelli , Reality is Not What it Seems, Allen Lane 2016, pp.229-231

24. Alexander Matthews in 'Lazarus Revived' (Firefly Books, 2019) p. 113.

25. Cassini Project - online: type in ' Lawsuit Almost Stalled Nasa's Cassini mission. 'In the one in ten chance of a mislaunch 73 lbs. of plutonium would have been scattered over Florida, rendering it uninhabitable for centuries. The launch went ahead anyway so that nearby planets could be examined more closely. See 'Cassini - NASA's Millennial Nightmare' by Chris Bryson, pp.2-3.

26. 'Alexander Matthews "How Some Scientists Erode the Human Rights We Value " , International Journal of Human Rights, Vol. 4, no.2 (Summer 2000),PUBLISHED BY FRANK CASS, LONDON pp.73-4.

27. Sellafield , ibid., p.74

28. Van Allen Belt , ibid. pp.76-78.

29. See Alexander Matthews, How Some Scientists Erode the Human Rights We Value, The International Journal of Human Rights, Vol. 4, No. 2, Summer 2000, pp67-78.)

30. W.B.Yeats, 'The Second Coming', p.235 'W.B.Yeats, The Poems', Everyman Library, 1992

31. Comment in The Times, Saturday 26 September 2020.

32. People and Power Report. www.purepla.net

33. Charles Olson, 'The Kingfisher', in 'Archaeologist of Morning', Cape Goliard Press, 1970 (pages unnumbered).

34. Soren Kierkegaard, 'Works of Love', p.105. Harper Perennial, 2009

35. Soren Kierkegaard, 'Purity of Heart, p.95. Fontana Books 1966

36. Bob Dylan, from the album, 'Slow Train Coming', 1979

37. Lynn White, Jr, 'Science, Vol 155, No 3767 (March 1967)

38. Ibid. p.53

39. Ibid. p.52

40. Michael J. Sandel, 'The Tyranny of Merit', Allen Lane, 2020

41. Ibid.p.41

42. Ibid.p.46

43. 'Intrapsychic': 'referring to processes occurring within the mind'. Critical Dictionary of Psychoanalysis', p.86. Penguin Reference, Charles Rycroft, 1995

44. For instance, for a discussion of the notion of essence with respect to the emotions. See L.F.Barrett, 'How Emotions are Made' p.164ff, Pan Books, 2017

45. Marcel Proust, II. Within A Budding Grove, p.456. Chatto 1992

46. Ludwig Wittgenstein, 'Tractatus Logico-Philosophicus', p.7. Routledge, 1966

47. See: C. Geertz, 'The Interpretation of Cultures', Basic Books, 1973; G. Samuel, 'Mind, body and culture',

Cambridge1990; M. Tomasello, 'Becoming Human', Harvard, 2019

48. I. McGilchrist, 'The Master and his Emissary', Yale, 2009

49. T.R. Blakeslee, 'The Right Brain', MacMillan Press, 1980; R. Joseph, 'The Right Brain and the Unconscious', Plenum, 1992. See also D.N. Stern in 'The Interpersonal World of the Infant', Basic Books, 2000 for the way the right hemisphere, or affectivity, is most developed in early infancy with the left hemisphere and language skills developing later. Also the role of language development on self-alienation (p.174). Tomasello also discusses the emergence of an 'objective' reality, and values, through peer interaction in childhood in 'Becoming Human', Harvard, 2019.

50. See D. Bohm, 'Wholeness and the Implicate Order', Chapters 1,2,3. D.N. Stern ('Interpersonal World..'), who says of the infant that it has no 'invariant pattern of awareness' (p.7), so there is no invariant organising of

experience. This is the 'raw' ground which becomes organised by the infant as it develops and categorised with the process of acculturation. For the Chinese/Taoist-Chan view see, D. Hinton, 'China Root', 'Awakened Cosmos', and 'Existence: A Story' (All published by Shambhala). There is a tendency to assume that there is only one way that a world can be put together and that this is reflected in a 'picture-theory of truth' (see Wittgenstein's 'Tractatus'), a view he subsequently rejected (see 'Philosophical Investigations' p31, Blackwell, 19667). This is manifestly not the case, and it is a mistake to assume that the worldview that one is immersed in, in our case the scientific-materialist, is the one and only one that is in some sense 'True' and corresponds to a reality as it is. See also E. Von Glaserfeld, 'An Introduction to Radical Constructivism' in 'The Invented Reality' ed. P. Watzlawick, Norton, 1984, who says:

'A brick builder who builds exclusively with bricks must sooner or later come to the conclusion that wherever there is to be an opening for a door or window, he has to make an arch to support the wall above. If this bricklayer then believes he has discovered a law of an absolute world, he makes much the same mistake as Kant when he came to believe the all geometry had to be Euclidean. Whatever we choose as building blocks, be it bricks or Euclidean elements, determines limiting constraints. We experience these constraints from the "inside", as it were, from the brick or Euclidean perspective. We never get to see the constraints of the world, with which our enterprises collide. What we experience, cognise, and come to know is necessarily built up of our own building blocks and can be explained in no other way than in terms of our ways and means of building.' (P.37)

See also, I. Berlin, 'Three Critics of the Enlightenment: Vico, Herder, and Hamann')

51. Daniel N. Stern 'The Interpersonal World of the Infant": It is during the second year of life that the use of language emerges and with it a new range of capacities, including cognitive ones (p.162ff), prior to that it is affectivity which is primarily operational.

52. D.N. Stern, 'The Interpersonal World the Infant', Basic Books 1985

53. See: neweconomics.org for instance.

54. See: R. Wilkinson and K. Pickett, 'The Spirit Level',Penguin, 2010 and D. Dorling, 'Injustice', Policy Press, 2011.

55. Statista: www.statista.com

56. See note 48 for references.

AUTHORS

Alexander Matthews was born in New York City in 1942. He taught Philosophy in a number of universities between 1975-1989. In 1986 he was awarded a visiting fellowship to Princeton University.

His book *A Diagram of Definition* on the Philosophy of Language was published in 1997 and *Lazarus Revived: An Atheist Argument For Conscious Life After Death* in 2019 by Firefly Books. His other writing includes three full length dramatic poems: *The Chairman* (1966), *Mr Swettham* (1969), and *Current Affairs* (1971), a book of short stories, *Brother To Sister* (1972), *Human Physics* (humour- 1973), a travel book, *A Traveller's Maze* (1974) and four poetic dramas, *Screaming Secrets (Inglewood Press, 2001)*. *Glass Roots* (2003), DO YOU LOVE THIS PLANET (2013), and MY ONE TRUE FRIEND Austin Macauley, 2020). He has also published several articles on Human Rights, including: *Philosophy and Human Rights*(1997), and *How Some Scientists Erode the Human Rights we Value* (2000) both in 'The International Journal of Human Rights', and *The Universe Has No Beginning? Doubts About The Big Bang Theory* (in Physics Essays 2006). He is currently working on two plays and a book of philosophy papers.

Gordon Ellis was Senior Lecturer in Communication Studies at the University of Salford, and has taught for the Open University, Extra-Mural Department of Manchester University, and Didsbury College of Education. He trained in, and taught, Philosophy, and has a particular interest in European and Eastern Philosophy and Psychology, as well as the Arts and Western Culture. For eight years he was Spiritual Director of Khandro Ling Tibetan Buddhist Centre in Cheshire.

His works include: *Crossing the Ocean of Existence* (2004), *Cycling Round Existence (Discovering Aspect of the Psychopathology of Western Civilisation* (2014), *Gordian Knots: Essays 2010-2020,* (2020), *Elysian Wiles: Poems 1990-2010, The Brittle Shards: Poems 2010-2019, At Times End:* Poem/s -2020, *Promontory* (Poem/s - 2021), and *Shades of Non-Existence, Sculpture of the Void* (Poems, 2021).

www.ingramcontent.com/pod-product-compliance
Lightning Source LLC
Chambersburg PA
CBHW071417210526
45465CB00001B/433